Student Booklet
Level E

Volume 2
Basic Facts

Photo Credits
©iStock International Inc., cover.

Acknowledgements
Content Consultant:

Linda Proudfit, Ph.D.

After earning a B.A. and M.A in Mathematics from the University of Northern Iowa, Linda Proudfit taught junior- and senior-high mathematics in Iowa. Following this, she earned a Ph.D. in Mathematics Education from Indiana University. She currently is Coordinator of Elementary Education and Professor of Mathematics Education at Governors State University in University Park, IL.

Dr. Proudfit has made numerous presentations at professional meetings at the local, state, and national levels. Her main research interests are problem solving and algebraic thinking.

www.WrightGroup.com

Copyright © 2009 by Wright Group/McGraw-Hill.

All rights reserved. Except as permitted under the United States Copyright Act, no part of this publication may be reproduced or distributed in any form or by any means, or stored in a database or retrieval system, without the prior written permission from the publisher, unless otherwise indicated.

Printed in USA.

Send all inquiries to:
Wright Group/McGraw-Hill
P.O. Box 812960
Chicago, IL 60681

ISBN 978-1-40-4568013
MHID 1-40-4568018

2 3 4 5 6 7 8 9 RHR 13 12 11 10 09

Contents

Tutorial Chart vii

Volume 2 Basic Facts
Topic 4 Addition and Subtraction Facts
Topic 4 Introduction 1
Lesson 4-1 Add or Subtract to Solve Problems 2–4
Lesson 4-2 Properties of Addition 5–7
Lesson 4-3 Addition Strategies 8–10
Lesson 4-4 Add Three 1-Digit Numbers 11–13
Lesson 4-5 Relate Addition and Subtraction 14–16
Lesson 4-6 Subtraction Strategies 17–19
Lesson 4-7 Fact Families 20–22
Topic 4 Summary 23
Topic 4 Mixed Review 24

Topic 5 Multiplication Facts
Topic 5 Introduction 25
Lesson 5-1 Meaning of Multiplication 26–28
Lesson 5-2 Multiply by 2, 5, and 10 29–31
Lesson 5-3 Properties of Multiplication 32–34
Lesson 5-4 Multiplication Strategies 35–37
Lesson 5-5 Basic Multiplication Facts 38–40
Topic 5 Summary 41
Topic 5 Mixed Review 42

Topic 6 Division Facts

Topic 6 Introduction 43
Lesson 6-1 Meaning of Division 44–46
Lesson 6-2 Properties of Zero and One 47–49
Lesson 6-3 Divide by 2, 3, 4, or 5 50–52
Lesson 6-4 Divide by 6, 7, 8, or 9 53–55
Lesson 6-5 Relate Multiplication and Division 56–58
Topic 6 Summary 59
Topic 6 Mixed Review 60

Glossary 61

Word Bank 62

Index 64

Objectives

Volume 2 Basic Facts

Topic 4 Addition and Subtraction Facts

Lesson	Objective	Pages
Topic 4 Introduction	**4.3** Use different thinking strategies to add numbers. **4.5** Illustrate the relationship between addition and subtraction. **4.6** Use different thinking strategies to subtract numbers.	1
Lesson 4-1 Add or Subtract to Solve Problems	**4.1** Solve problems by choosing the operation and show the meanings of addition and subtraction.	2–4
Lesson 4-2 Properties of Addition	**4.2** Apply various addition properties including the commutative, associative, and identity properties.	5–7
Lesson 4-3 Addition Strategies	**4.3** Use different thinking strategies to add numbers.	8–10
Lesson 4-4 Add Three 1-Digit Numbers	**4.4** Find the sum of three 1-digit numbers.	11–13
Lesson 4-5 Relate Addition and Subtraction	**4.5** Illustrate the relationship between addition and subtraction.	14–16
Lesson 4-6 Subtraction Strategies	**4.6** Use different thinking strategies to subtract numbers.	17–19
Lesson 4-7 Fact Families	**4.7** Write addition and subtraction fact families.	20–22
Topic 4 Summary	Review skills related to addition and subtraction facts.	23
Mixed Review 4	Maintain concepts and skills.	24

Topic 5 Multiplication Facts

Lesson	Objective	Pages
Topic 5 Introduction	**5.1** Use repeated addition, arrays, and counting by multiples to do multiplication. **5.3** Recognize and use the properties of zero and one and the commutative and associative properties of multiplication. **5.5** Memorize to automaticity the multiplication table for numbers 1 through 10.	25
Lesson 5-1 Meaning of Multiplication	**5.1** Use repeated addition, arrays, and counting by multiples to do multiplication.	26–28
Lesson 5-2 Multiply by 2, 5, and 10	**5.2** Know the multiplication tables of 2s, 5s, and 10s (to "times 10").	29–31
Lesson 5-3 Properties of Multiplication	**5.3** Recognize and use the basic properties of multiplication.	32–34

Lesson	Objective	Pages
Lesson 5-4 Multiplication Strategies	**5.4** Apply multiplication strategies such as skip counting and doubles.	35–37
Lesson 5-5 Basic Multiplication Facts	**5.5** Memorize to automaticity the multiplication table for numbers 1 through 10.	38–40
Topic 5 Summary	Review multiplication concepts and skills.	41
Mixed Review 5	Maintain concepts and skills.	42

Topic 6 Division Facts

Lesson	Objective	Pages
Topic 6 Introduction	**6.1** Define division by showing the relationship between multiplication facts and division facts. **6.2** Understand the special properties of 0 and 1 in multiplication and division. **6.3** Show the patterns of dividing by 2, 3, 4, or 5. **6.5** Show and apply the families of facts for multiplication and division.	43
Lesson 6-1 Meaning of Division	**6.1** Use models and multiplication to understand division.	44–46
Lesson 6-2 Properties of Zero and One	**6.2** Understand the special properties of 0 and 1 in multiplication and division.	47–49
Lesson 6-3 Divide by 2, 3, 4, or 5	**6.3** Find the basic facts involving division by 2, 3, 4, or 5.	50–52
Lesson 6-4 Divide by 6, 7, 8, or 9	**6.4** Find the basic facts involving division by 6, 7, 8, or 9.	53–55
Lesson 6-5 Relate Multiplication and Division	**6.5** Write multiplication and division fact families.	56–58
Topic 6 Summary	Review division skills.	59
Mixed Review 6	Maintain concepts and skills.	60

Tutorial Guide

Each of the standards listed below has at least one animated tutorial for students to use with the lesson that matches the objective. If you are using the electronic components of *Pinpoint Math*, you will find a complete listing of Tutorial codes and titles when you access them either online or via CD-ROM.

Level E

Standards by topic	Tutorial codes
Volume 2 Basic Facts	
Topic 4 Addition and Subtraction Facts	
4.1 Solve problems by choosing the operation and show the meanings of addition and subtraction.	4a Using Models, to Show the Meaning of Addition and Subtraction
4.1 Solve problems by choosing the operation and show the meanings of addition and subtraction.	4b Using Addition Fact Strategies
4.1 Solve problems by choosing the operation and show the meanings of addition and subtraction.	4c Using Subtraction Fact Strategies
4.1 Solve problems by choosing the operation and show the meanings of addition and subtraction.	4d Choosing the Operation to Solve Addition and Subtraction Word Problems
4.2 Apply various addition properties including the commutative, associative, and identity properties.	4f Using Fact Families to Add and Subtract
4.2 Apply various addition properties including the commutative, associative, and identity properties.	4g The Commutative and Associative Properties of Addition
4.3 Use different thinking strategies to add numbers.	4h Understanding Compatible Numbers
4.3 Use different thinking strategies to add numbers.	4i Finding the Sum of Three Numbers
4.3 Use different thinking strategies to add numbers.	4j Using Addition Fact Strategies
4.3 Use different thinking strategies to add numbers.	4d Choosing the Operation to Solve Addition and Subtraction Word Problems
4.4 Find the sum of three 1-digit numbers.	4i Finding the Sum of Three Numbers
4.5 Illustrate the relationship between addition and subtraction.	4f Using Fact Families to Add and Subtract
4.5 Illustrate the relationship between addition and subtraction.	4c Using Subtraction Fact Strategies
4.6 Use different thinking strategies to subtract numbrs.	4f Using Fact Families to Add and Subtract
4.6 Use different thinking strategies to subtract numbrs.	4c Using Subtraction Fact Strategies
4.7 Write addition and subtraction fact families.	4f Using Fact Families to Add and Subtract
Topic 5 Multiplication Facts	
5.1 Use repeated addition, arrays, and counting by multiples to do multiplication.	5a Understanding Multiplication, Example A
5.1 Use repeated addition, arrays, and counting by multiples to do multiplication.	5b Understanding Multiplication, Example B
5.3 Recognize and use the basic properties of multiplication.	5c Using the Commutative and Associative Properties of Multiplication
5.3 Recognize and use the basic properties of multiplication.	5a Understanding Multiplication, Example A
5.3 Recognize and use the basic properties of multiplication.	5b Understanding Multiplication, Example B
5.3 Recognize and use the basic properties of multiplication.	5d Using Patterns to Solve Word Problems
5.5 Memorize to automatically the multiplication table for numbers 1 through 10.	5e Using Fact Families to Multiply and Divide

Topic 6 Division Concepts	
6.1 Use models and multiplication to understand division.	6a Understanding Division
6.1 Use models and multiplication to understand division.	6b Modeling Division
6.3 Find the basic facts involving division by 2, 3, 4, or 5.	6c Understanding Patterns for Division
6.4 Find the basic facts involving division by 6, 7, 8, or 9.	6c Understanding Patterns for Division
6.5 Write multiplication and division fact families.	6d Using Fact Families to Multiply and Divide

Topic 4: Addition and Subtraction

Topic Introduction

Complete with teacher help if needed.

1.
   a. Add _____ counters to make 10.

   b. List the ways you could separate the 6 counters into 2 groups. Write the addition fact that goes with each.

   1 and _____   1 + 5 = 6

   2 and _____   _____

   3 and _____   _____

   Objective 4.3: Use different thinking strategies to add numbers.

2. Complete the steps.

   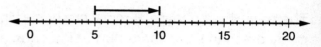

   Start at 5.

   Move _____ to the _____.

   End at _____.

   Objective 4.3: Use different thinking strategies to add numbers.

3. Use this model to write a subtraction fact.

   ○⊗⊗⊗

   4 − 3 = _____

   Objective 4.6: Use different thinking strategies to subtract numbers.

4. Use this model to write an addition fact.

   3 + 3 = _____

   Now use the model to write a subtraction fact.

   _____ − _____ = _____

   Objective 4.5: Illustrate the relationship between addition and subtraction.

Lesson 4-1 — Add or Subtract to Solve Problems

Model It

Words to Know **Addition** means joining groups or increasing a quantity.
Subtraction means taking away, comparing, or decreasing a quantity.

Activity 1

There are 5 elephants. 8 more are born. How many elephants are there now? Use counters to show the meaning of 5 + 8.

Join 5 to 8. There are 13 in all.

Practice 1

Karen has 9 books. She gets 3 for her birthday. How many books does she have now? Use counters to show the meaning of 9 + 3.

Join _____ to _____.

There are _____ in all.

Activity 2

Carl has 13 grapes. He eats 8. How many grapes are left? Use counters to show 13 − 8. Then use words to show the meaning of 13 − 8.

Take 8 away from 13. There are 5 left.

Practice 2

There are 12 crows in a tree. 3 fly away. How many are left? Use counters to show 12 − 3. Use the words below to show the meaning of 12 − 3.

Take _____ away from _____.

There are _____ left.

On Your Own

Kim has 12 bracelets. Her sister Angela gives her 3 more. Do you add or subtract to find how many bracelets Kim has?

_____

Write About It

Saul has 9 games. He gives 4 to Val. How do you find how many Saul has left?

Objective 4.1: Solve problems by choosing the operation and show the meanings of addition and subtraction.

Lesson 4-1: Add or Subtract to Solve Problems

B — Understand It

Example 1

It costs $11 to go to the water park. It costs $5 to go to the zoo. How much more does it cost to go to the water park?
Use counters to compare 11 and 5.

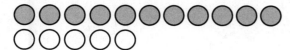

Cross off the counters that match.

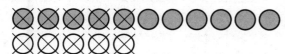

The difference is 6. $11 - 5 = 6$

Practice 1

There are 7 kinds of sandwiches in the lunchroom. There are 2 kinds of pizza. How many more kinds of sandwiches are there? Use counters to compare 7 and 2.

The difference is _____.

_____ − _____ = _____

Example 2

There were 13 dogs at the park. Then 6 left. How many are there now? Use a number line to show this problem.

Start at 13. Move 6 to the left. You stop at 7.

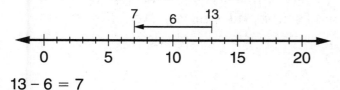

$13 - 6 = 7$

Practice 2

There were 6 piano students. Then 7 more students joined the class. How many are there now? Use a number line to show this problem.

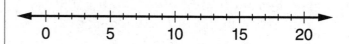

Start at _____. Move _____ to the _____.

Where do you stop? _____

What is the fact? _____

On Your Own

Jack's school is 12 miles from his home. The library is 4 miles closer. Do you add or subtract to find the distance from Jack's home to the library? Use a number line.

_____

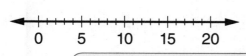

Write About It

Write a comparison word problem to go with this model.

Objective 4.1: Solve problems by choosing the operation and show the meanings of addition and subtraction.

Lesson 4-1 — Add or Subtract to Solve Problems

1. Use this model to write an addition and a subtraction fact.

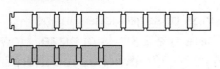

_____

_____

2. There are 12 monkeys at the zoo. Eight are adults. How many are **not** adults? Draw a number line model.

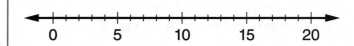

3. Dan has 12 yards of rope, but he needs 20 yards. How can he find out how much more rope he needs to buy? Circle the letter of the correct answer.

A Add 12 to 20. **B** Subtract 12 from 20.

C Add 20 to 12. **D** Subtract 20 from 12.

4. Lani has 7 yards of fabric. She buys 9 more yards. How many yards does she have now? Draw a number line model.

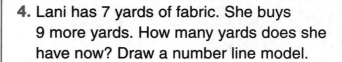

5. Sue is 15 years old. Bob is 6 years younger than Sue. Show the number fact you use to figure out how old Bob is.

_____

_____

6. Darrell has 8 fish. Kari has 5 fish. What problem can you use to find how many fish they have all together? Circle the letter of the correct answer.

A $8 - 5$ **B** $8 + 5$

C $5 - 8$ **D** 85

7. Jake baked 12 muffins. Then Janine gave him 4 more. What operation do you use to find how many muffins Jake has all together? What fact can you write for this problem?

_____

_____

8. Explain the difference between these phrases.

8 decreased by 5   8 increased by 5

_____

_____

Objective 4.1: Solve problems by choosing the operation and show the meanings of addition and subtraction.

| Lesson 4-2 | Properties of Addition |

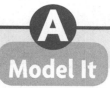

Words to Know — **Commutative property of addition:** add numbers in any order and the sum will be the same.
  For example, 9 + 7 = 16 and 7 + 9 = 16.
Associative property of addition: group numbers in any way and the sum will be the same.
  For example, (4 + 5) + 8 = 17 and 4 + (5 + 8) = 17.

Activity 1

Use MathFlaps to show
8 + 6 = 6 + 8.

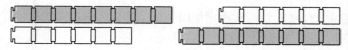

8 + 6 = 14
6 + 8 = 14

Practice 1

Use MathFlaps to show
9 + 5 = 5 + 9.

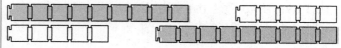

5 + 9 = _____

9 + 5 = _____

Activity 2

Use MathFlaps to show
(4 + 3) + 5 = 4 + (3 + 5).

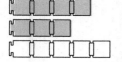 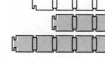

(4 + 3) + 5 = 7 + 5 = 12
4 + (3 + 5) = 4 + 8 = 12

Practice 2

Use MathFlaps to show
(1 + 4) + 3 = 1 + (4 + 3).

 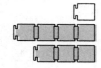

(1 + 4) + 3 = _____ + 3 = _____

1 + (4 + 3) = 1 + _____ = _____

On Your Own

Write the missing number.

(10 + 6) + 8 = 10 + (_____ + 8)

Write About It

You know that the sum of 5 and 8 is 13. Explain how you can find the sum of 8 and 5.

Objective 4.2: Apply various addition properties including the commutative, associative, and identity properties.

Lesson 4-2 | Properties of Addition

Understand It

Words to Know — The **identity property of addition** states that if you add 0 to any number, the sum is the number itself.

Example 1

Add 6 + 0.
Model with MathFlaps.
Start with 6 MathFlaps.

Add 0 MathFlaps. You still have 6 MathFlaps. You can also use the identity property of addition.

6 + 0 = 6

Practice 1

Add 0 + 7.
Model with MathFlaps.

Start with _____ MathFlaps.

Add _____ MathFlaps.

You have _____ MathFlaps.

You can also use the identity property of addition.

0 + 7 = _____

Example 2

The properties of addition can be used to add mentally.

3 + 0 + 7

= 3 + 7 + 0   commutative property

= 10 + 0   associative property

= 10   identity property

Practice 2

Use the properties of addition to help add.
0 + 6 + 4

= 6 + _____ + 0   _____ property

= _____ + _____   _____ property

= _____   _____ property

On Your Own

Use the properties to add the following.

5 + 0 + 3 + 5

Write About It

What number makes the sentence
5 + 7 = _____ + 5 true? Explain how you know.

Objective 4.2: Apply various addition properties including the commutative, associative, and identity properties.

Lesson 4-2 Properties of Addition

1. Shade the MathFlaps in two ways to show $5 + 3 = 3 + 5$.

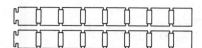

2. Write the missing number.

   a. $4 + 9 = 9 +$ _____

   b. $6 + ($_____$+ 7) = (6 + 8) + 7$

   c. _____ $+ 8 = 8$

   d. $2 + (5 + 3) = ($_____$+ 5) + 3$

   e. $1 + (7 + 9) = 1 + (9 +$_____$)$

3. Identify the property used. Write C for commutative, A for associative, or I for identity.

   a. $8 + 3 = 3 + 8$ _____

   b. $0 + 3 = 3$ _____

   c. $(9 + 7) + 5 = 9 + (7 + 5)$ _____

   d. $(8 + 3) + 6 = (3 + 8) + 6$ _____

4. Which sentence shows the associative property? Circle the letter of the correct answer.

   A $14 + 24 = 24 + 14$

   B $26 + (18 + 9) = 26 + (9 + 18)$

   C $(12 + 16) + 4 = 12 + (16 + 4)$

   D $(6 + 25) = (25 + 6)$

5. Use the properties to add.
   $0 + 1 + 2 + 3 + 4 + 5 + 6 + 7 + 8 + 9$

6. Explain how you can add $3{,}407 + 0$.

7. Add from left to right. Then use the properties to check your work.
   $6 + 7 + 4 + 3 + 2 + 8$

8. Which property is used in this number sentence? $(3 + 5) + 8 = (5 + 3) + 8$

Objective 4.2: Apply various addition properties including the commutative, associative, and identity properties.

Lesson 4-3 Addition Strategies

Words to Know   A **number line** is a line that shows numbers from least to greatest.

Activity 1

7 + ? = 10
Use MathFlaps to show 7.
Compare to a row of 10 MathFlaps.
How many more do you need?

7 + 3 = 10

Practice 1

Use MathFlaps to make 10.

4 + _____ = 10

8 + _____ = 10

_____ + 9 = 10

_____ + 5 = 10

Activity 2

6 + 5 = ?

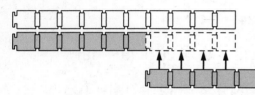

Compare 6, the greater number, to 10.
I need 4 more to make 10.
I can move 4 from the smaller number
to the 6 to make 10.

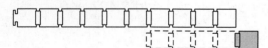

The new addition fact is 10 + 1 = 11.
So 6 + 5 = 11.

Practice 2

8 + 6 = ?
Use MathFlaps to model.

The greater number is _____.

I need _____ more to make 10.

I can move _____ from the smaller number to the 8 to make 10.

The new addition fact is 10 + _____ = _____.

So 8 + 6 = _____.

On Your Own

Make 10 to find the sum.

8 + 4 = _____

9 + 7 = _____

7 + 6 = _____

Write About It

Explain how to make a 10 when adding 9 + 7. Then find the sum of 9 + 7.

Objective 4.3: Use different thinking strategies to add numbers.

Lesson 4-3: Addition Strategies

B Understand It

Example 1

Count on to find 3 + 12.

Start at the greater number, 12.

Count on 3 more numbers: 13, 14, 15

3 + 12 = 15

Practice 1

Count on to find 14 + 4.

The greater number is _____.

Count on 4 more numbers: _____, _____, _____, _____

14 + 4 = _____

Example 2

Use a doubles fact to find the sum.

If 5 + 5 = 10, then 5 + **6** = 11.
↑
this is one more

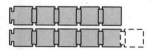

If 10 + 10 = 20, then 10 + 11 = 21.

Practice 2

Use a doubles fact to find the sum.

If 11 + 11 = _____, then 11 + 12 = _____.

If 8 + 8 = _____, then 8 + 9 = _____.

On Your Own

Find the sum.
Which strategy did you use?

6 + 7 = _____ _____

13 + 2 = _____ _____

8 + 7 = _____ _____

Write About It

Which doubles fact would you use to find 7 + 8? Explain.

Objective 4.3: Use different thinking strategies to add numbers.

Lesson 4-3: Addition Strategies

1. Find the number needed to make 10.

  a. 4 + _____ = 10

  b. 2 + _____ = 10

  c. 3 + _____ = 10

2. Make 10 to solve. Write the "plus 10" problem.

  a. 8 + 7 _____
  _____

  b. 9 + 4 = _____
  _____

  c. 6 + 5 = _____
  _____

3. Count on to solve. Show how you counted.

  a. 2 + 7 = _____
  Count on: _____

  b. 11 + 4 = _____
  Count on: _____

  c. 3 + 4 = _____
  Count on: _____

4. Use a doubles fact to solve. Write the doubles fact you used.

  a. 8 + 9 _____
  _____

  b. 5 + 4 = _____
  _____

  c. 7 + 8 = _____
  _____

5. Jen downloaded 9 songs on Monday. She downloaded 7 songs on Tuesday. How many songs did Jen download in all? Write a fact to solve.

  _____ + _____ = _____

  Which strategy did you use? _____

6. Explain how to find 8 + 5.

7. Which of the following has a sum of 14?

  A 8 + 8   B 1 + 4

  C 4 + 4   D 9 + 5

8. Cory has 7 pencils. How many more pencils does Cory need to make 10?

  A 3   B 4

  C 17   D 13

Objective 4.3: Use different thinking strategies to add numbers.

Lesson 4-4: Add Three 1-Digit Numbers

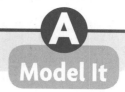

Activity 1

Find the sum: 1 + 5 + 6.
Add two numbers at a time.
Use MathFlaps to check your answer.

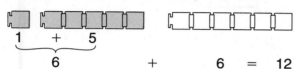

1 + 5
6   +   6 = 12

Practice 1

Find the sum: 3 + 2 + 9.
Use MathFlaps to check your answer.

3 + 2 = _____

5 + 9 = _____

3 + 2 + 9 = _____

Activity 2

Add 7 + 5 + 3.
Add the numbers in any order. Use MathFlaps to check your answer.

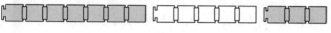

10 + 5 = 15

You can add 7 and 3 mentally.
Their sum, 10, is easy to add to another number.

Practice 2

Add 9 + 6 + 1.
Add the numbers in any order. Use MathFlaps to check your answer.

Which two numbers did you add first?

_____

9 + 6 + 1 = _____

On Your Own

Austin has 5 red T-shirts, 3 white T-shirts, and 5 blue T-shirts. How many does he have in all?

5 + 5 = _____

10 + 3 = _____

5 + 3 + 5 = _____

Austin has _____ T-shirts.

Write About It

Cindy says that you must always add numbers in the order they are given in the problem. Sarah disagrees. Who is right? Explain.

Objective 4.4: Find the sum of three one-digit numbers.

Lesson 4-4: Add Three 1-Digit Numbers

B Understand It

Example 1

Add 6 + 3 + 4.
First try to make 10 with two of the numbers.
Then add the third number to that sum of 10.

6 + 3 + 4
↓ ↙
6 + 4 = 10

10 + 3 = 13

6 + 3 + 4 = 13

Practice 1

Add 9 + 1 + 8.
First try to make 10 with two of the numbers.

_____ + _____ = 10

10 + _____ = _____

9 + 1 + 8 = _____

Example 2

5 roses, 8 carnations, and 9 daisies are displayed in 3 vases. How many flowers are in the vases?

Write the number sentence for the story.

5 + 8 + 9 = ?

Add. 5 + 8 + 9 = 22

The vases have 22 flowers.

Practice 2

Jasmine is on the track team. She ran 2 miles on Monday, 5 miles on Tuesday, and 6 miles on Friday. How far did she run on those 3 days?

Complete the number sentence.

2 + _____ + _____ = _____

Jasmine ran _____ miles.

On Your Own

Matt plays basketball. He made 6 baskets in each game for 2 games in a row and 7 baskets in the third game. How many baskets did he make in all?

_____

Write About It

What numbers would you add first to make finding the sum of 5 + 9 + 5 easier? Explain.

Objective 4.4: Find the sum of three one-digit numbers.

Lesson 4-4: Add Three 1-Digit Numbers

Try It

1. Find the sum. Write the addition sentence for the picture.

☐☐☐ + ■■■ + ☐☐☐☐

_____

2. Add.

a. $5 + 2 + 4 =$ _____

b. $10 + 3 + 3 =$ _____

c. $4 + 2 + 6 =$ _____

d. $5 + 8 + 1 =$ _____

3. Find these sums.

a. $2 + 5 + 8 =$ _____

b. $7 + 7 + 7 =$ _____

c. $6 + 9 + 2 =$ _____

d. $8 + 6 + 6 =$ _____

4. In which problem could you **not** make 10 when adding?

A $3 + 7 + 1$ B $9 + 1 + 6$

C $4 + 9 + 2$ D $4 + 8 + 2$

5. At an art show, Room 1 has 6 projects. Room 2 has 3 projects. Room 3 has as many projects as Room 1. How many projects are in the 3 rooms?

_____

6. Ryan is playing a game. He rolls three number cubes and looks at the numbers—1, 2, 3, 4, 5, or 6—on top. Then he finds the sum of the three numbers. What is the greatest total he can get?

_____

7. Tiffany adds $4 + 7 + 2$. She says $4 + 7 = 10$ and $10 + 2$ is 12, so the sum is 12. Is she right?

8. Celeste tossed 3 number cubes with 1, 2, 3, 4, 5, and 6 on each. The sum of the numbers she rolled is less than 15. Could she have rolled two 6s? Explain.

Objective 4.4: Find the sum of three one-digit numbers.

Lesson 4-5: Relate Addition and Subtraction

Activity 1

There used to be 12 circles. Some were taken away. Now there are 7. How many were taken away?

Draw more circles until there are 12.

Complete the number sentences.
7 + 5 = 12   and   12 − 5 = 7

Practice 1

There used to be 14 squares. Some were taken away. Now there are 6. How many were taken away?

Draw more squares until there are 14. Complete the number sentences.

6 + _____ = 14   and   14 − _____ = 6

Activity 2

13 − 8 = ?

Model the problem. Use addition to help.

Separate the counters. Put 8 in one group.

How many are in the other group? Think: 8 plus how much is 13? 8 + **5** = 13

There are 5 in the other group, so
13 − 8 = **5**.

Practice 2

12 − 8 = ?

Model the problem.

Separate the counters.

What addition fact do the counters show?

8 + 4 = _____

12 − 8 = _____

On Your Own

What addition fact can help you solve 13 − 7?

_____ + _____ = _____

Write About It

How can you choose the right addition fact to help you to solve a subtraction problem?

Objective 4.5: Illustrate the relationship between addition and subtraction.

| Lesson 4-5 | **Relate Addition and Subtraction** | **B Understand It** |

Example 1

15 − 7 = ?

Write an addition fact to help you solve.

Think: 7 plus how much is 15?

7 + **8** = 15

So, 15 − 7 = **8**.

Practice 1

20 − 11 = ?

Write an addition fact to help you solve.

Think: 11 plus how much is 20?

11 + _____ = 20

So, 20 − 11 = _____.

Example 2

Miguel read 18 pages and Thomas read 15 pages. How many more pages did Miguel read?

Write an addition and subtraction sentence to model the problem.

15 + **3** = 18

18 − 15 = **3**

Miguel read **3** more pages than Thomas.

Practice 2

There are 4 cans of soup in the pantry. There used to be 9 cans. How many are missing?

Write an addition and subtraction sentence to model the problem.

4 + _____ = 9

9 − 4 = _____

There are _____ missing cans.

On Your Own

Monique rode her bike for 15 minutes. Stefan rode his bike for 11 minutes. Write addition and subtraction sentences to show how many more minutes Monique rode her bike.

_____

_____

She rode _____ minutes longer.

Write About It

Explain why the addition sentence 5 + ? = 11 is also like a subtraction problem.

Objective 4.5: Illustrate the relationship between addition and subtraction.

Lesson 4-5: Relate Addition and Subtraction

Try It

1. There used to be 11 triangles. Now there are 4.

   △ △ △ △

   Draw more triangles until there are 11.

   Complete the number sentences.

   4 + _____ = 11

   11 − 4 = _____

2. Margie scored 7 points and Henry scored 9 points. To find how many more points Henry scored than Margie, which number sentence could be used?

   A 9 + 7 = 16   B 16 − 9 = 7

   C 9 − 7 = 2   D 7 − 9 = 2

3. Write a subtraction problem that can help you solve 2 + _____ = 11.

   _____ − _____ = _____

4. Write an addition fact to help you solve 12 − 3. Then solve.

   3 + _____ = 12

   12 − 3 = _____

5. Fill in the blank to make the number sentence true.

   17 − _____ = 10

6. Write an addition fact that will help you solve 20 − 5.

   5 + _____ = _____

   20 − 5 = _____

7. Melanie has 12 pencils now. She used to have 18. How many are missing? Write an addition and subtraction sentence to model the problem. Then solve.

   12 + _____ = _____

   _____ − 12 = _____

   _____ pencils are missing.

8. Rosa knows that 4 + 5 = 9. How will this help her find the difference of 9 − 5?

Objective 4.5: Illustrate the relationship between addition and subtraction.

Lesson 4-6: Subtraction Strategies

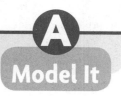
Model It

Activity 1

Use counters to find the difference.

7 − 3

○○○○⊗⊗⊗

Start with 7 counters. Take 3 away. 4 are left.

7 − 3 = 4

Practice 1

Use counters to find the difference.

11 − 5

○○○○○○○○○○○

11 − 5 = _____

Activity 2

Count back to find the difference.
9 − 2

Start at 9. Count back two numbers:
8, 7

9 − 2 = 7

Practice 2

Count back to find the difference.
12 − 3

Start at _____

Count back _____ numbers:

_____, _____, _____

12 − 3 = _____

On Your Own

Find the difference.

15 − 4 = _____

8 − 5 = _____

12 − 6 = _____

Write About It

Draw a diagram to show how to find 12 − 5. Explain how you got your answer.

Objective 4.6: Use different thinking strategies to subtract numbers.

Lesson 4-6 Subtraction Strategies

Understand It

Words to Know An **addend** is a number that is added to another number.

Example 1

Use the addition problems to find the difference.

a. [2]   8
   + 6   − 6
   ———   ———
    8   [2]

b. [4]   14
   +10   −10
   ———   ———
    14   [4]

Practice 1

Write the missing addend, and then find the difference.

a.   13
   +10   −10
   ———   ———
    13

b.   12
   + 7   − 7
   ———   ———
    12

c.   17
   +11   −11
   ———   ———
    17

Example 2

Write the addition fact to help find the difference.

a. 14   [6]
   − 8   +8
   ———   ———
   [6]   14

b.   9   [4]
   − 5   +5
   ———   ———
   [4]   9

Practice 2

Write the addition fact to help find the difference.

a.   9
   − 8   +
   ———   ———

b. 15
   − 4   +
   ———   ———

c.   7
   − 2   +
   ———   ———

On Your Own

Write the missing number. Use addition to check.

a. 9 − 3 = _____

b. 13 − 9 = _____

Write About It

Solve 18 − 11. Explain how you can use an addition fact to check the difference.

Objective 4.6: Use different thinking strategies to subtract numbers.

Lesson 4-6: Subtraction Strategies

Try It

1. Find the difference.

   a. 8 − 2 = _____

   b. 18 − 10 = _____

   c. 13 − 4 = _____

2. What subtraction fact is shown?

   ○○○○○⊗⊗⊗

3. Joe wants to find the difference of 17 and 3 by counting back. Explain how Joe can do this.

4. Which subtraction fact is related to the addition fact 7 + 2 = 9?

   A 9 − 7 = 2

   B 11 − 2 = 9

   C 16 − 9 = 7

   D 7 − 2 = 5

5. Write the missing number.

   a. 18 − 9 = _____

   b. 8 − 0 = _____

   c. 12 − 9 = _____

6. Write the addition fact to help find the difference.

   7 − 3 = _____   _____ + _____ = _____

   16 − 10 = _____   _____ + _____ = _____

7. Explain how you can use addition to solve 7 − 3.

8. Find the difference.

   a. 20 − 4 = _____

   b. 16 − 8 = _____

   c. 17 − 3 = _____

Objective 4.6: Use different thinking strategies to subtract numbers.

Volume 2 — 19 — Level E

Lesson 4-7: Fact Families

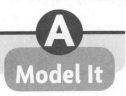

Activity 1

Find the fact family shown with these MathFlaps.

8 + 3 = 11
3 + 8 = 11
11 − 8 = 3
11 − 3 = 8

Practice 1

Find the fact family shown with these MathFlaps.

____ + ____ = ____

____ + ____ = ____

____ − ____ = ____

____ − ____ = ____

Activity 2

Sketch MathFlaps to show the addition fact 6 + 4 = 10.

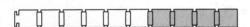

Write the rest of the fact family.
6 + 4 = 10
4 + 6 = 10
10 − 4 = 6
10 − 6 = 4

Practice 2

Sketch MathFlaps to show the subtraction fact 13 − 4 = 9.

Write the rest of the fact family.
13 − 4 = 9

____ − ____ = ____

____ + ____ = ____

____ + ____ = ____

On Your Own

Brett has 18 counters.
He gives 10 counters to Ashley.
What number fact does this show?
Write the rest of the fact family.

_____

_____

Write About It

If you know an addition fact, explain how you can find the whole fact family.

Objective 4.7: Write addition and subtraction fact families.

Lesson 4-7 — Fact Families

B Understand It

Example 1

Write the fact family for 6, 7, and 13.
Circle the total in each fact.

6 + 7 = 13
7 + 6 = 13
13 − 6 = 7
13 − 7 = 6

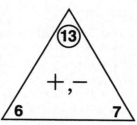

Practice 1

Write the sum of 7 and 8 in the circle.
Complete the fact family.

7 + 8 = ◯

_____ + _____ = ◯

◯ − _____ = _____

◯ − _____ = _____

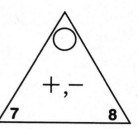

Example 2

Use a related fact to solve 12 − 5.

Start with the total and subtract.
12 − 5 = _____
12 − _____ = 5

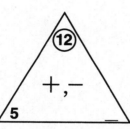

Add. The total should be the sum.
_____ + 5 = 12   5 + _____ = 12

Now complete one fact.
7 + 5 = 12, so 7 is the missing number.
12 − 5 = 7

Practice 2

Use a related fact to solve 13 − 8.

Start with the total and subtract.
_____
_____

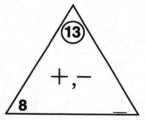

Add. The total should be the sum.
_____   _____

Now complete one fact.
Write the missing number in the triangle.

13 − 8 = _____

On Your Own

Complete the triangle.
Write the fact family.

_____

_____

_____

_____

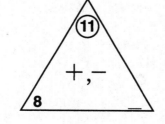

Write About It

Explain how to use the sum in a family of facts.

Objective 4.7: Write addition and subtraction fact families.

Lesson 4-7   **Fact Families**

1. Write two addition facts and two subtraction facts for the model.

   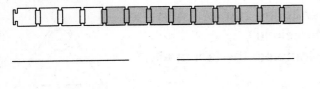

   _____   _____

   _____   _____

2. Complete the fact family by filling in the blanks.

12 + _____ = 18	_____ + 12 = 18
18 − 12 = _____	18 − _____ = 12

3. Draw a fact triangle you could use to solve 12 − 3. Then draw a fact triangle you could use to solve 12 + 3.

4. Look back at Exercise 3. How are the triangles alike? How are they different?

5. Write all the addition and subtraction facts that use the numbers 6, 8, and 14.

   _____

   _____

6. Name the other subtraction sentence that is in the same family as 15 − 3 = 12.

   _____

7. Write the fact family that has the numbers 16, 8, and 8.

   _____

   _____

8. What belongs to the same fact family as 10 + 2 = 12? Circle the letter of the correct answer.

   A 12 + 2 = 14   **B** 12 − 10 = 2

   C 10 − 2 = 8   **D** 12 + 10 = 22

Objective 4.7: Write addition and subtraction fact families.

Topic 4: Addition and Subtraction Facts

Topic Summary

Choose the correct answer. Explain how you decided.

1. Tiffany has a collection of postcards from family vacations. If she collects 5 more postcards on the next vacation she will have a total of 14 postcards. How many postcards are in her collection now?

   A 5   **B** 14   **C** 19   **D** 9

   _____
   _____
   _____

2. Which of these does **not** belong to the same fact family as 6 + 3 = 9?

   A 9 − 6 = 3   **B** 3 + 6 = 9   **C** 9 − 3 = 6   **D** 6 − 3 = 3

   _____
   _____
   _____

Objective: Review skills related to addition and subtraction facts.

Topic 4: Addition and Subtraction Facts

Mixed Review

1. Write the number in expanded notation.

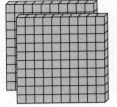

_____

_____

Volume 1, Lesson 3-2

2. Find each sum.

a. 7 + 3 + 6 = _____

b. 8 + 5 + 6 = _____

c. 7 + 4 + 2 = _____

d. 9 + 5 + 1 = _____

Volume 2, Lesson 4-4

3. Round each number to the nearest thousand and ten thousand.

a. 271,106 _____

b. 97,485 _____

c. 183,719 _____

d. 5,249,607 _____

Volume 1, Lesson 3-4

4. What is the greatest whole number that makes the sentence true? Explain.

10 + ___ < 20

_____

_____

Volume 1, Lesson 1-3

5. What is the largest number that rounds to 1,200?

_____

Volume 1, Lesson 3-4

6. There were some carrots in a dish. Jamie and 3 friends each ate 2 carrots. There were 16 carrots left. How many carrots were on the plate originally?

_____

Volume 2, Lesson 4-2

Objective: Maintain concepts and skills.

Topic 5: Multiplication Facts

Topic Introduction

Complete with teacher help if needed.

1.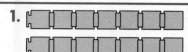

  a. Find the sum.

  _____ + _____ = _____

  b. Find the product.

  _____ × _____ = _____

2.

  a. Find the sum.

  ___ + ___ + ___ + ___ = ___

  b. Find the product.

  _____ × _____ = _____

Objective 5.1: Use repeated addition, arrays, and counting by multiples to do multiplication.

3. Multiply.

  a. 6 × 0 = _____

  b. 6 × 1 = _____

  c. 8 × 1 = _____

  d. 8 × 0 = _____

4. Multiply.

  a. 3 × 6 = _____

  b. 8 × 3 = _____

  c. 9 × 4 = _____

  d. 7 × 5 = _____

Objective 5.3: Recognize and use the properties of zero and one and the commutative and associative properties of multiplication.

Objective 5.5: Memorize to automaticity the multiplication table for numbers 1 through 10.

Lesson 5-1: Meaning of Multiplication

Words to Know Multiplication is a way to join groups of equal size. You can use **repeated addition** to multiply. Add the same number as many times as needed.

Activity 1

Use MathFlaps to show 3 groups of 5.

Find 3 × 5 using repeated addition.

5 + 5 + 5 = 15

Practice 1

Use MathFlaps to show 2 groups of 7.

Find 2 × 7 using repeated addition.

_____ + _____ = _____

Activity 2

Shade 3 rows of 6 squares on grid paper.

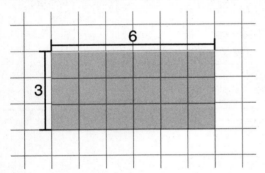

Find 3 × 6 using repeated addition.

6 + 6 + 6 = 18

Practice 2

Shade 4 rows of 8 squares on grid paper.

Find 4 × 8 using repeated addition.

8 + 8 + 8 + 8 = _____

On Your Own

Draw 2 rows of 8 dots to show 2 × 8. What two numbers can you add to find 2 × 8? Write the addition fact.

Write About It

Carlos covered his kitchen floor with square tiles. Without counting, how can he find the number of tiles he used?

Objective 5.1: Use repeated addition, arrays, and counting by multiples to do multiplication.

Lesson 5-1: Meaning of Multiplication

B Understand It

Words to Know You can make jumps on a number line, or **skip count**, to multiply.

Example 1

Find 6 × 4 by skip counting by 4s.

Make 6 jumps of 4 on a number line.

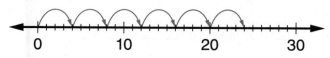

4, 8, 12, 16, 20, 24

6 × 4 = 24

Practice 1

Find 5 × 3 by skip counting by 3s. Start at 0.

Make 5 jumps of 3 on a number line.

_____, _____, _____, _____, _____

5 × 3 = _____

Example 2

Find 5 × 5 by skip counting by 5s. Use the nickels to help you count.

5, 10, 15, 20, 25

5 × 5 = 25

Practice 2

Find 6 × 10 by skip counting by 10s. Use the dimes to help you count.

_____, _____, _____, _____, _____, _____

6 × 10 = _____

On Your Own

Pamela biked 5 miles per hour for 6 hours. How far did she bike? Count by 5s to get the answer. Write the multiplication sentence.

_____

_____

Write About It

Isabel skip counted by 3s to solve 3 × 9. Veena skip counted by 9s. Whose way do you think is better? Explain, and show how that way works.

Objective 5.1: Use repeated addition, arrays, and counting by multiples to do multiplication.

Lesson 5-1: Meaning of Multiplication

Try It

1. Shade rows of squares to make a model for 4×6. Write the fact below the model.

2. Draw a number line model for 9×3. Write the fact below the model.

   0   10   20   30

   _____

3. Write a multiplication fact for this model.

   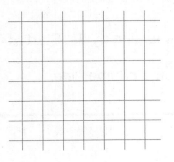

   _____

4. Which has the same meaning as 6×5?

   A $6 \times 6 \times 6 \times 6 \times 6$

   B $6 + 6 + 6 + 6 + 6 + 6$

   C $5 + 5 + 5 + 5 + 5 + 5$

   D $5 + 5 + 5 + 5 + 5$

5. What are two different ways you can add to solve 7×4?

   _____

   _____

6. Here is a strategy for multiplying by 9.
   Try 9×5.
   10 fives = 50
   9 fives is 1 five fewer than 10 fives,
   so 9 fives = 50 − 5
   $9 \times 5 = 45$
   Use this strategy to find 9×7.

   _____

   _____

   _____

Objective 5.1: Use repeated addition, arrays, and counting by multiples to do multiplication.

Lesson 5-2: Multiply by 2, 5, and 10

A Model It

Words to Know — You can use a **multiplication facts table** to find products. For example, to find 5 × 8, trace across the row for 5 until you are in the column for 8.

×	0	1	2	3	4	5	6	7	8	9	10
2	0	2	4	6	8	10	12	14	16	18	20
5	0	5	10	15	20	25	30	35	40	45	50
10	0	10	20	30	40	50	60	70	80	90	100

Activity 1

Use the row for 2. Look in this row to find these facts.

2 × 3 = 6   2 × 5 = 10
2 × 7 = 14   2 × 9 = 18

Practice 1

Use the row for 5. Look in this row to complete these facts.

5 × 7 = _____   5 × 6 = _____

5 × 5 = _____   5 × 4 = _____

Activity 2

Use the pattern in the row for 10 to find 10 × 8.
In the row for 10, the ones digit is always 0. To multiply by 10 quickly, add a zero to the number you are multiplying.

10 × 8 = 80

Practice 2

Multiply by 10. Use the table to check your answers.

10 × 9 = _____   10 × 7 = _____

10 × 5 = _____   10 × 3 = _____

On Your Own

Multiply.

2 × 8 = _____   9 × 5 = _____

4 × 10 = _____   2 × 10 = _____

Write About It

Why does the number 20 appear three times in the table?

Objective 5.2: Know the multiplication tables of 2s, 5s, and 10s (to "times 10").

| Lesson 5-2 | **Multiply by 2, 5, and 10** | **B Understand It** |

Words to Know Numbers you multiply are called **factors**. The total you find when you multiply is the **product**.

16 is the product of 2 and 8.
2 and 8 are factors of 16.

$2 \times 8 = 16$
factor factor product

Example 1

What factor makes this sentence true?

$2 \times 8 = 8 \times$ _____

Changing the order of the factors does not change the product.

$2 \times 8 = 16$ $8 \times 2 = 16$
Both products are 16, so $2 \times 8 = 8 \times 2$

Practice 1

What factor makes this sentence true?

$10 \times 7 = 7 \times$ _____

What are the products? _____

Example 2

Write two multiplication facts using the factors 5 and 9.

Multiply the factors to find the product. Switch the order of the factors to get the other fact.

$5 \times 9 = 45$ $9 \times 5 = 45$

Practice 2

Write two multiplication facts using the factors 4 and 10.

_____ $\times$ _____ = _____

_____ $\times$ _____ = _____

On Your Own

Find the products.

$10 \times 5 =$ _____ $4 \times 2 =$ _____

$5 \times 8 =$ _____ $6 \times 10 =$ _____

$8 \times 5 =$ _____ $2 \times 7 =$ _____

Write About It

Axel says that if $9 \times 2 = 18$, then $18 \times 2 = 9$. What is his mistake?

Objective 5.2: Know the multiplication tables of 2s, 5s, and 10s (to "times 10").

Lesson 5-2 — **Multiply by 2, 5, and 10**

Try It

1. Write two multiplication facts using the factors 6 and 5.

   ____ × ____ = ____

   ____ × ____ = ____

2. Multiply each of these numbers by 2.

   a. 8 ____   b. 9 ____   c. 4 ____

   d. 3 ____   e. 7 ____   f. 1 ____

3. Find the products.

   a. $5 \times 8 =$ ____   b. $2 \times 5 =$ ____

   c. $10 \times 4 =$ ____   d. $6 \times 5 =$ ____

   e. $5 \times 1 =$ ____   f. $10 \times 8 =$ ____

   g. $7 \times 10 =$ ____   h. $2 \times 9 =$ ____

4. $9 \times 5 =$

   A 14

   B 45

   C 54

   D 95

5. When you multiply a number by 5, what is the ones digit in the product?

   _____

6. Fill in the missing number.

   $4 \times 10 = 10 \times$ ____

7. Explain how counting by 5s can help you find 7×5.

8. Think about the numbers from 10 to 20. Which numbers are products you can make when you use 2 as a factor? Write the multiplication facts for these products.

   _____

   _____

   _____

   _____

Objective 5.2: Know the multiplication tables of 2s, 5s, and 10s (to "times 10").

Volume 2   Level E

Lesson 5-3: Properties of Multiplication

Words to Know

Commutative property of multiplication: You can multiply factors in any order, and the product will be the same.
 For example, $3 \times 8 = 24$ and $8 \times 3 = 24$.

Associative property of multiplication: You can group factors in any way, and the product will be the same.
 For example, $(5 \times 4) \times 6 = 120$ and $5 \times (4 \times 6) = 120$.

Activity 1

Draw a diagram to show $4 \times 6 = 6 \times 4$.

$4 \times 6 = 24$ $6 \times 4 = 24$
So, $4 \times 6 = 6 \times 4$.

Practice 1

Draw a diagram to show that $3 \times 5 = 5 \times 3$.

$3 \times 5 =$ _____ $5 \times 3 =$ _____

So, $3 \times 5 =$ _____ $\times$ _____.

Activity 2

Use the commutative and associative properties to find $5 \times 9 \times 2$.

$5 \times 9 \times 2$

$= 5 \times 2 \times 9$ commutative property

$= (5 \times 2) \times 9$ associative property

$= 10 \times 9 = 90$

Practice 2

Use the commutative and associative properties to find $2 \times 10 \times 2$.

$2 \times 10 \times 2$

$= 2 \times$ _____ $\times 10$ _____

$= ($ _____ $\times$ _____ $) \times 10$ _____

$=$ _____ $\times 10 =$ _____

On Your Own

Write the missing number.

$(3 \times 9) \times 4 =$ _____ $\times (9 \times 4)$

Write About It

Explain how you know that $2 \times 4 \times 6$ is the same as 4×12.

Objective 5.3: Recognize and use the basic properties of multiplication.

Lesson 5-3: Properties of Multiplication

Understand It (B)

Example 1

Use MathFlaps to find 5 × 1.

5 × 1 = 5

Practice 1

Use MathFlaps to find 1 × 7.

1 × 7 = _____

Example 2

Use a diagram to solve this problem. Kris gained 0 pounds every week for 8 weeks. How much did she gain or lose?

Day 1	Day 2	Day 3	Day 4
0	0	0	0

Day 5	Day 6	Day 7	Day 8
0	0	0	0

0 × 8 = 0 + 0 + 0 + 0 + 0 + 0 + 0 + 0 = 0

She did not gain or lose any weight.

Practice 2

Use a diagram to solve this problem. Aaron saw 0 movies every day for 6 days. How many movies did he see in all?

He saw _____ movies.

On Your Own

a. Draw a model to solve 1 × 12.

1 × 12 = _____

b. Use repeated addition to solve 0 × 12.

0 × 12 = _____

Write About It

Explain what the product will be if you multiply a number by 1, and what the product will be if you multiply by 0.

Objective 5.3: Recognize and use the basic properties of multiplication.

Lesson 5-3: Properties of Multiplication

1. Draw a diagram to show that $6 \times 3 = 3 \times 6$.

2. Write the missing number.

a. $12 \times 6 = 6 \times$ _____

b. $9 \times ($ _____ $\times 8) = (9 \times 4) \times 8$

c. _____ $\times 12 = 12 \times 5$

d. $12 \times (9 \times 5) = ($ _____ $\times 9) \times 5$

e. $35 \times (12 \times 6) = 35 \times (6 \times$ _____$)$

3. Identify the property used. Write C for commutative, A for associative, or B for both.

a. $5 \times (4 \times 2) = (5 \times 4) \times 2$ _____

b. $15 \times 9 \times 4 = (15 \times 4) \times 9$ _____

c. $15 \times 74 = 74 \times 15$ _____

d. $(17 \times 6) \times 6 = 17 \times (6 \times 6)$ _____

4. Which sentence shows the commutative property of multiplication?

A $18 \times 9 \times 5 = 18 \times 5 \times 9$

B $24 \times 1 = 12 \times 2 \times 1$

C $(13 \times 9) \times 6 = 13 \times (9 \times 6)$

D $(4 + 13) + 8 = 4 + (13 + 8)$

5. Multiply.

$12 \times 5 \times 1 \times 2 \times 0$

6. Use the commutative and associative properties to simplify the calculation.

$2 \times 4 \times 5 \times 10$

7. Multiply from left to right. Then use the properties to check your work.

$2 \times 8 \times 5 \times 5 \times 2 \times 1$

8. Write the rule for multiplying by 1. Give an example.

Objective 5.3: Recognize and use the basic properties of multiplication.

Lesson 5-4: Multiplication Strategies

Words to Know   The **multiples** of 4 are the products of any counting number and 4.

Activity 1

Use the chart to find 4 × 6.

1	2	3	4	5	6	7	8	9	10
11	12	13	14	15	16	17	18	19	20
21	22	23	24	25	26	27	28	29	30

Start at 4. Circle 4. Count on four numbers to find the next multiple of 4. Circle the number. Keep going until you have six circles.

1	2	3	(4)	5	6	7	(8)	9	10
11	(12)	13	14	15	(16)	17	18	19	(20)
21	22	23	(24)	25	26	27	28	29	30

What is the sixth number circled? 24

What is 4 × 6? 24

Practice 1

Use the chart to find 5 × 3.

1	2	3	4	5	6	7	8	9	10
11	12	13	14	15	16	17	18	19	20
21	22	23	24	25	26	27	28	29	30

Start at 5.

1	2	3	4	5	6	7	8	9	10
11	12	13	14	15	16	17	18	19	20
21	22	23	24	25	26	27	28	29	30

Circle 5. Count on five numbers to find the next multiple of 5. Circle the number.
Keep going until you have three circles.

What is the third number circled? _____

What is 5 × 3? _____

Activity 2

Skip count to find 3 × 6.

Count by 3s. Say six numbers.

3, 6, 9, 12, 15, 18

3 × 6 = 18

Practice 2

Skip count to find 5 × 4.

Count by _____s. Say _____ numbers.

5, 10, _____, _____

5 × 4 = _____

On Your Own

Skip count to find 8 × 3.

_____

8 × 3 = _____

Write About It

To multiply by 9, Maria counts on 10, and then goes back 1.
Start at **9**. Think "10 more makes 19." Go back 1 to **18**. Think "10 more makes 28." Go back 1 to **27**.
Explain how to find two more numbers. Then tell what problem you solved.

Objective 5.4: Apply multiplication strategies such as skip counting and doubles.

Lesson 5-4 — Multiplication Strategies

Words to Know   When you **double** a number, you add it to itself.

Example 1

Find 6 × 8.
Draw an array 6 units wide and 8 units long.

Cut the array in half.
Add the products of the two smaller arrays.

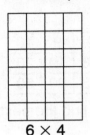 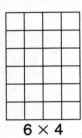

6 × 4   6 × 4

6 × 4 = 24   6 × 4 = 24
       24 + 24 = 48, so 6 × 8 = 48.

Practice 1

Find 7 × 6.
On a piece of grid paper, draw an array 7 units wide and 6 units long.

Cut the array in half.
Add the products of the two smaller arrays.

7 × 3 = _____   7 × 3 = _____

_____ + _____ = _____,

so 7 × 6 = _____.

Example 2

Cut a factor in half and double the product to find 4 × 12.

4 × 12 is the same as two sets of 4 × 6.

Find 4 × 6.
4, 8, 12, 16, 20, 24
4 × 6 = 24

Double 24.
24 + 24 = 48
4 × 12 = 48

Practice 2

Cut a factor in half and double the product to find 6 × 8.

6 × 8 is the same as two sets of 3 × _____.

Find _____.
_____
_____

Double _____.
_____ + _____ = _____

6 × 8 = _____

On Your Own

Six friends were playing marbles. Each of them had 12 marbles. How many marbles were there in all? Cut a factor in half and double to find the answer.

Write About It

Selena knows that 8 × 4 = 32. How could she use that fact to find 8 × 8?

Objective 5.4: Apply multiplication strategies such as skip counting and doubles.

Lesson 5-4: Multiplication Strategies

1. Skip count to find 2×9.

   2, 4, _____

   $2 \times 9 =$ _____

2. Use a hundred chart to find 6×9.

   _____

   $6 \times 9 =$ _____

3. Jordan wants to find 9×8. Which multiplication fact can he double to find the answer?

   A $9 \times 4 = 36$ **B** $4 \times 8 = 32$

   C $5 \times 8 = 40$ **D** $9 \times 16 = 144$

4. Kelli's chorus room has 8 rows of chairs. There are 7 chairs in each row. How many chairs are there in all? Show how to skip count to find the answer.

   _____

   _____

   _____

5. Use the arrays to solve 9×8.

   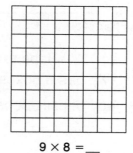

   $9 \times \_\_ = \_\_$ $9 \times \_\_ = \_\_$ $9 \times 8 = \_\_$

   $9 \times 8 =$ _____

6. Cut a factor in half and double the product to solve.

   a. $8 \times 4 =$ _____

   b. $8 \times 5 =$ _____

   c. $12 \times 8 =$ _____

7. Isaiah has 8 packs of plastic spiders. Each pack has 5 spiders. How many spiders does he have all together?

   _____

8. Achala and Clay are making oatmeal cookies for a bake sale. They can make 12 cookies with 1 cup of oats. How many cookies can they make if they have 4 cups of oats?

Objective 5.4: Apply multiplication strategies such as skip counting and doubles.

Lesson 5-5: Basic Multiplication Facts

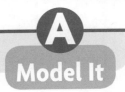
Model It

Words to Know   A **product** is the answer to a multiplication problem.
The numbers being multiplied are **factors**.
An **array** is a picture that shows a multiplication fact.

Activity 1

Use MathFlaps. Find the product of 3 × 6.

First, find how many MathFlaps there are in 2 rows.
2 × 6 = 6 + 6 = 12
Add one more row of 6.
3 × 6 = 12 + 6 = 18

Practice 1

Use MathFlaps to find the product of 4 × 8.

How many rows? _____

How many in each row? _____

4 × 8 = ___ + ___ + ___ + ___ = ___

Activity 2

Use MathFlaps to find 6 × 4.

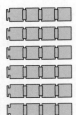

The top three rows show 3 × 4 = 12.
The bottom three rows show 3 × 4 = 12.
So, 6 × 4 = 12 + 12 = 24.

Practice 2

Use MathFlaps to solve this problem. Kayla has 8 boxes of 10 pencils each. How many pencils does she have? _____

On Your Own

Draw an array to show 9 × 8. Find the product.

9 × 8 = _____

Write About It

What multiplication fact could you use to help you find 6 × 7? How could this fact be used?

Objective 5.5: Memorize to automaticity the multiplication table for numbers between 1 and 10.

Lesson 5-5 — Basic Multiplication Facts

B Understand It

Words to Know

The rows and columns in the **multiplication facts table** can help you find the facts.

$$6 \times 8 = 48$$
factor factor product

×	0	1	2	3	4	5	6	7	8	9	10
0	0	0	0	0	0	0	0	0	0	0	0
1	0	1	2	3	4	5	6	7	8	9	10
2	0	2	4	6	8	10	12	14	16	18	20
3	0	3	6	9	12	15	18	21	24	27	30
4	0	4	8	12	16	20	24	28	32	36	40
5	0	5	10	15	20	25	30	35	40	45	50
6	0	6	12	18	24	30	36	42	48	54	60
7	0	7	12	21	28	35	42	49	56	63	70
8	0	8	16	24	32	40	48	56	64	72	80
9	0	9	18	27	36	45	54	63	72	81	90
10	0	10	20	30	40	50	60	70	80	90	100

Example

Find the missing factor.
$5 \times \underline{} = 45$

Find the "5" row. Look along the row to the right to find 45. Then look at the top of the column to find the missing factor.
$5 \times 9 = 45$

Practice

Find the missing factors.

1. $9 \times \underline{} = 72$

2. $\underline{} \times 8 = 56$

3. $3 \times \underline{} = 27$

4. $\underline{} \times 6 = 48$

On Your Own

Write two different multiplication problems that have different factors with a product of 24.

_____

_____

Write About It

Look at the facts for the even numbers 2, 4, 6, 8, 10. What can you say about the products?

Objective 5.5: Memorize to automaticity the multiplication table for numbers between 1 and 10.

Lesson 5-5   **Basic Multiplication Facts**

1. Write a multiplication fact for the array.

   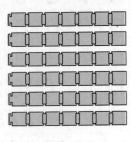

   _____

2. Multiply.

   a. $9 \times 8 =$ _____

   b. $7 \times 8 =$ _____

   c. $4 \times 9 =$ _____

   d. $6 \times 7 =$ _____

3. Find the missing factor.

   a. $6 \times$ _____ $= 30$

   b. $8 \times$ _____ $= 48$

   c. $4 \times$ _____ $= 36$

   d. $8 \times$ _____ $= 64$

4. Which has a product of 60? Circle the letter of the correct answer.

   A 5×10

   B 10×6

   C 9×6

   D 9×7

5. Write two multiplication facts using the factors 6 and 9.

   _____

6. Put these facts in order from greatest product to least.

   4×9 8×7

   6×5 3×9

   _____

   _____

7. Maeve says 8×8 is 32 because $4 \times 4 = 16$, and 8 is twice 4. Do you agree? Explain.

   _____

8. Describe two different ways to find the product of 7×6.

   _____

Objective 5.5: Memorize to automaticity the multiplication table for numbers between 1 and 10.

Topic 5: Multiplication Facts

Topic Summary

Circle the letter of the correct answer. Explain how you decided.

1. Marianne earned $9 each hour for 6 hours. How much did she earn in all?

   A $9

   B $60

   C $54

   D $15

   _____
   _____
   _____

2. Kevin gave 1 book to each of the 15 friends who came to his birthday party. How many books did he give to his friends?

   A 30

   B 15

   C 1

   D 0

   _____
   _____
   _____

Objective: Review skills related to multiplication facts.

Topic 5: Multiplication Facts

Mixed Review

1. Find each sum.

   a. 2 + 8 = _____   b. 7 + 8 = _____

   c. 9 + 7 = _____   d. 3 + 2 = _____

   e. 2 + 5 = _____   f. 3 + 8 = _____

2. Write <, >, or =.

   a. 310 ◯ 299

   b. 209 ◯ 290

   c. 452 ◯ 425

   d. 54 ◯ 540

   Volume 2, Lesson 4-3   *Volume 1, Lesson 2-4*

3. What number added to 6 gives 11?

   _____

4. Find the missing factor:

   4 × _____ = 36

   Volume 2, Lesson 4-5   *Volume 2, Lesson 5-5*

5. Jose knows 9 × 8 = 72. How can he use that fact to find 8 × 9?

   _____

   _____

   _____

6. Madison says 29 × 38 × 0 × 3 × 1 = 1 because the number is multiplied by 1. Is she correct? Explain.

   _____

   _____

   _____

   Volume 2, Lesson 5-3   *Volume 2, Lesson 5-3*

Objective: Maintain concepts and skills.

Topic 6: Division Facts

Topic Introduction

Complete with teacher help if needed.

1. Write the facts.

   a. Write a multiplication fact.

   _____ × _____ = _____

   b. Write a division fact.

   _____ ÷ _____ = _____

   Objective 6.1: Use models and multiplication to understand division.

2. Divide.

   a. 6 ÷ 1 = _____

   b. 0 ÷ 4 = _____

   Objective 6.2: Understand the special properties of 0 and 1 in multiplication and division.

3. Divide.

   a. 16 ÷ 2 = _____

   b. 27 ÷ 3 = _____

   c. 35 ÷ 5 = _____

   d. 24 ÷ 4 = _____

   Objective 6.3: Show the patterns of dividing by 2, 3, 4, or 5.

4. Complete the fact family.

   15 ÷ 3 = 5

   a. _____ ÷ _____ = _____

   b. _____ × _____ = _____

   c. _____ × _____ = _____

   Objective 6.5: Show and apply the families of facts for multiplication and division.

Volume 2 — Level E

Lesson 6-1 | **Meaning of Division**

Words to Know

```
      4  ← quotient
   2)‾8
   ↑     ↑
divisor  dividend
```

Activity 1

Use MathFlaps. Find 8 ÷ 2.
Connect 8 flaps.

Fold the MathFlaps to make 2 stacks.

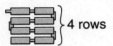 } 4 rows

Count the rows. There are 4.
8 ÷ 2 = 4

Practice 1

Use MathFlaps. Find 18 ÷ 3.

How many stacks? _____

How many rows? _____

18 ÷ 3 = _____

Activity 2

Sunelle has 32 flowers for 8 gift baskets. How many flowers can she put in each basket?

Start with the total to be divided, 32.

Divide in 8 equal groups.

32 ÷ 8 = 4

She can put 4 flowers in each gift basket.

Practice 2

Thad has 27 soccer balls to pack in 9 boxes. How many balls will he pack in each box?

The total to be divided is _____.

Divide in _____ equal groups.

_____

He will pack _____ soccer balls in each box.

On Your Own

Divide 284 by 4.

284 ÷ 4 = _____

Write About It

Describe a MathFlaps model of 45 ÷ 9.

Objective 6.1: Use models and multiplication to understand division.

Lesson 6-1: Meaning of Division

B Understand It

Example 1

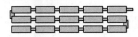

Write a division fact and a multiplication fact to describe the model.

To divide, start with the total, 15.
Divide by the number of rows, 3.
How many are in each row? 5
$15 \div 3 = 5$

To multiply, start with the number in each row, 5.
Multiply by the number of rows, 3.
The product is the total, 15.
$5 \times 3 = 15$

Practice 1

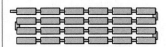

Write a division fact and a multiplication fact to describe the model.

Total: _____

Number of rows: _____

Number in each row: _____

Divide. _____

Multiply. _____

Example 2

$10 \div 2 = ?$
Use multiplication to help you divide.
$? \times 2 = 10$
$5 \times 2 = 10$
$10 \div 2 = 5$ because $5 \times 2 = 10$.

Check: Divide 10 MathFlaps into 2 groups.

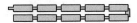

Practice 2

$30 \div 6 = ?$
Draw a picture and use multiplication to help.

$30 \div 6 =$ _____ because _____ $\times 6 = 30$.

On Your Own

Find the missing numbers.

_____ $\times 3 = 24$ $24 \div 3 =$ _____

_____ $\times 3 = 9$ $9 \div 3 =$ _____

_____ $\times 9 = 36$ $36 \div 9 =$ _____

Write About It

Explain how to use multiplication facts to help find the answer to division problems.

Objective 6.1: Use models and multiplication to understand division.

Lesson 6-1: Meaning of Division

1. Find the quotient. Write a multiplication sentence to help.

   $56 \div 7 = ?$

   _____

   $56 \div 7 =$ _____

2. Find the quotient.

   a. $45 \div 9 =$ _____

   b. $32 \div 4 =$ _____

   c. $64 \div 8 =$ _____

   d. $30 \div 3 =$ _____

3. Find the product or quotient.

   a. $30 \div 6$ _____

   b. _____ $\times 6 = 30$

   c. $42 \div 7 =$ _____

   d. _____ $\times 7 = 42$

4. Which multiplication fact can you use to help find $48 \div 6$?

   A $48 \times 6 = 288$ C $48 \div 8 = 6$

   B $12 \times 4 = 48$ D $8 \times 6 = 48$

5. Write a division problem you can solve if you know that $8 \times 5 = 40$.

   _____

6. Use multiplication to explain why 5 divided by 0 does not equal 0.

7. A waiter wants to earn $36 in tips during his 6-hour shift. How much does he have to earn each hour?

   _____

8. There are 7 empty darkrooms. The photography teacher has 21 students, and she wants to assign the same number to work in each darkroom. How many students will work in each darkroom?

   _____

Objective 6.1: Use models and multiplication to understand division.

Lesson 6-2: Properties of Zero and One

Activity 1

Draw a picture to find $0 \div 4$.

$0 \div 4 = 0$

Practice 1

Draw a picture. Find $0 \div 6$.

$0 \div 6 =$ _____

Activity 2

Find the problem that cannot be solved.

$5 \times 0 \quad 0 \times 5 \quad 0 \div 5 \quad 5 \div 0$

Both multiplication facts can mean 5 groups of 0, or $0 + 0 + 0 + 0 + 0$. $5 \times 0 = 0$ and $0 \times 5 = 0$.

Try $0 \div 5$. Start with 0 MathFlaps. Can you make 5 groups of 0? Yes, there will be 0 in each group, so $0 \div 5 = 0$.

Try $5 \div 0$. Start with 5 MathFlaps. Can you make 0 groups? No, you have to have at least 1 group. $5 \div 0$ cannot be solved.

Practice 2

Circle the problem that cannot be solved. Solve the other problems.

$10 \div 0 =$ _____ $10 \times 0 =$ _____

$0 \div 10 =$ _____ $0 \times 10 =$ _____

On Your Own

a. Draw a model to solve $0 \div 12$.

$0 \div 12 =$ _____

b. Draw a model to solve 0×12.

$0 \times 12 =$ _____

Write About It

Explain the difference between dividing 0 by a number, and dividing a number by 0.

Objective 6.2: Understand the special properties of 0 and 1 in multiplication and division.

Lesson 6-2: Properties of Zero and One

B Understand It

Example 1

Draw a diagram to find 6 ÷ 1.

6 ÷ 1 = 6

Practice 1

Draw a diagram to find 8 ÷ 1.

8 ÷ 1 = _____

Example 2

Compute: 2 × 7 × 1

2 × 7 × 1

14 × 1 = 14

2 × 7 × 1 = 14

Practice 2

Compute: 8 × 1 × 3

8 × 1 × 3 = _____

On Your Own

Take the product of 40 and 1. Divide it in 1 part. What is (40 × 1) ÷ 1?

Write About It

Compare what happens when you multiply a number by 1 and when you divide a number by 1.

Objective 6.2: Understand the special properties of 0 and 1 in multiplication and division.

Lesson 6-2 — **Properties of Zero and One**

1. Write the multiplication problems shown by this diagram.

   _____

2. Write *true* if the statement is correct. Write *false* if the statement is not correct.

   a. 6 × 1 = 1 _____

   b. 0 × 9 = 0 _____

   c. 15 ÷ 1 = 15 _____

   d. 0 ÷ 24 = 24 _____

3. Multiply or divide.

   a. 12 × 9 × 0 × 3 = _____

   b. (8 × 3) ÷ 1 = _____

   c. 0 ÷ (7 × 3) = _____

   d. 18 × (0 ÷ 12) = _____

4. What is 4 ÷ 1?

   A 0

   B 4

   C 3

   D 1

5. Circle the problem that cannot be solved. Solve the other problems.

   0 × 3 = _____   3 × 0 = _____

   0 ÷ 3 = _____   3 ÷ 0 = _____

6. Explain why 0 ÷ 6 = 0.

   _____

   _____

7. Write a rule to show what happens when you multiply a number by 1.

8. Jon says he watches 0 hours of television each day. How much television will he watch in 7 days? A month? A year? Explain.

Objective 6.2: Understand the special properties of 0 and 1 in multiplication and division.

Lesson 6-3 — **Divide by 2, 3, 4, or 5**

Words to Know   A **remainder** is a number left over after you divide.

Activity 1

a. To find 8 ÷ 2, start with 8 MathFlaps. Fold to make groups of 2.

There are 4 groups of 2.
8 ÷ 2 = 4

b. Now find 9 ÷ 2. Start with 9 MathFlaps. When you fold to make groups of 2, there is 1 MathFlap left over.

There are 4 groups. The remainder is 1.

9 ÷ 2 = 4 R1

Practice 1

Use MathFlaps to solve. Circle the problem that has no remainder.

a. 14 ÷ 2 = ?

Fold 14 MathFlaps to make groups of _____. There are _____ groups of _____, with _____ left over.

14 ÷ 2 = _____

b. 15 ÷ 2 = ?

Fold 15 MathFlaps to make groups of _____. There are _____ groups of _____, with _____ left over.

15 ÷ 2 = _____

Activity 2

Which of these can be divided evenly by 3: 15, 16? Write that division fact.

First try 15 ÷ 3. There are 5 equal groups, so **15 ÷ 3 = 5.**

Then try 16 ÷ 3. There is a remainder, so 16 ÷ 3 is not a basic fact.

Practice 2

Which of these can be divided evenly by 3: 8, 9, 10, 11, 12, 13, 14? Write those division facts.

Use MathFlaps.

Basic facts: _____, _____

On Your Own

From 2 through 11, which numbers can be divided evenly by 2? Use MathFlaps.

Write About It

How can you find all the "divided by 3" facts from 30 ÷ 3 through 3 ÷ 3?

Objective 6.3: Find the basic facts involving division by 2, 3, 4, or 5.

Lesson 6-3: Divide by 2, 3, 4, or 5

B Understand It

Words to Know A number is **divisible** by another number when there is no remainder after division.

Example 1

From 0 through 20, which numbers are divisible by 4? Write the basic facts.

Start with 4. $4 \div 4 = 1$, so 4 **is** divisible by 4.

Try the next number, 5. $5 \div 4 = 1\ R1$, so 5 **is not** divisible by 4.

Keep trying numbers. The only numbers that can be divided by 4 without leaving a remainder are 4, 8, 12, 16, and 20.

$4 \div 4 = 1$ $8 \div 4 = 2$ $12 \div 4 = 3$

$16 \div 4 = 4$ $20 \div 4 = 5$

Practice 1

From 21 through 40, which numbers are divisible by 4?

Start with 21. $21 \div 4 =$ _____

Try 22. $22 \div 4 =$ _____

Keep trying numbers.

Then write the basic facts.

_____

_____

_____

Example 2

Find the patterns in dividing by 5.

Dividing by 5

Dividend	Quotient	Remainder
5	1	None
6	1	1
7	1	2
8	1	3
9	1	4
10	2	None
11	2	1
12	2	2

Practice 2

Find the patterns in dividing by 5.

Dividing by 5

Dividend	Quotient	Remainder
13		3
14	2	
15		None
16		
17	3	
18		3
19	3	4
20		

On Your Own

Jasper's class has 32 students. Yasmin's class has 34 students. Which class could be divided in 4 equal groups?

Write About It

Use the numbers from 4 through 16 in order. Divide by 4. What is the pattern?

Objective 6.3: Find the basic facts involving division by 2, 3, 4, or 5.

Lesson 6-3: Divide by 2, 3, 4, or 5

Try It

1. What division problem does each set of MathFlaps show?

_____   _____

2. Find all the numbers in the list that are divisible by 2. Write the facts.

11 12 13 14 15 16 17 18 19 20

_____

_____

_____

3. Divide by 4. Fill in the table.

Dividing by 4

Dividend	Quotient	Remainder
20		None
21	5	
22		
23		3
24		

4. Divide by 3. Fill in the table.

Dividing by 3.

Dividend	Quotient	Remainder
12		
13		
14		
15		

5. Solve.

a. $30 \div 5 =$ _____

b. $40 \div 5 =$ _____

c. $25 \div 5 =$ _____

d. $15 \div 5 =$ _____

6. Jack has 32 tomato seeds. Lucien has 33 pumpkin seeds. Who will be able to plant all his seeds in 4 equal rows?

A Jack   B Lucien

C Both   D Neither

7. Mr. Washington had some pencils, and he gave equal numbers to 3 students. How many pencils might Mr. Washington have had?

8. When you are dividing, what is the greatest possible remainder? Explain.

Objective 6.3: Find the basic facts involving division by 2, 3, 4, or 5.

Lesson 6-4: Divide by 6, 7, 8, or 9

 Model It

Words to Know A **remainder** is a number left over after you divide.

Activity 1

a. To find 18 ÷ 6, start with 18 MathFlaps.

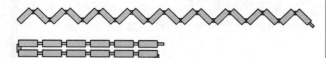

Fold to make groups of 6.
There are 3 groups of 6.
18 ÷ 6 = 3

b. Find 19 ÷ 6. Start with 19 MathFlaps. When you fold to make groups of 2, there is 1 MathFlap left over.

There are 3 groups. The remainder is 1.
19 ÷ 6 = 3 R1

Practice 1

Use MathFlaps to solve. Circle the problem that has no remainder.

a. 30 ÷ 6 = ?

Fold 30 MathFlaps to make groups of _____. There are _____ groups of _____, with _____ left over.

30 ÷ 6 = _____

b. 32 ÷ 6 = ?

Fold 32 MathFlaps to make groups of _____. There are _____ groups of _____, with _____ left over.

32 ÷ 6 = _____

Activity 2

Which of these can be divided evenly by 7: 20, 21? Write that division fact.

First try 20 ÷ 7. There is a remainder, so 16 ÷ 3 is not a basic fact.

Then try 21 ÷ 7. There are 3 equal groups, so **21 ÷ 7 = 3**.

Practice 2

Which of these can be divided evenly by 7: 27, 28, 29, 30, 31, 32, 33, 34, 35? Write those division facts.

Use MathFlaps.

Basic facts: _____, _____

On Your Own

From 6 through 36, which numbers can be divided evenly by 6? Use MathFlaps.

Write About It

42 is divisible by 7. How can you find the next whole number divisible by 7? Write the fact.

Objective 6.4: Find the basic facts involving division by 6, 7, 8, or 9.

Lesson 6-4 Divide by 6, 7, 8, or 9

Understand It

Words to Know A number is **divisible** by another number when there is no remainder after division.

Example 1

From 0 through 40, which numbers are divisible by 8? Write the basic facts.

Start with 8. $8 \div 8 = 1$

To find the next number, add 8 to the dividend. $16 \div 8 = 2$

Keep adding 8 to the dividend.

$24 \div 8 = 3$ $32 \div 8 = 4$ $40 \div 8 = 5$

Practice 1

From 41 through 80, which numbers are divisible by 8? Write the basic facts.

You know 40 is divisible by 8. Add 8 to find the next number. Keep adding 8.

_____

_____

_____

Example 2

Find the possible remainders you divide by 9.

Dividing by 9

Dividend	Quotient	Remainder
18	1	0
19	1	1
20	1	2
21	1	3
22	1	4
23	1	5
24	1	6
25	1	7
26	1	8
27	2	0

Practice 2

Find the rest of the "divided by 9" facts. Continue until you reach $90 \div 9$.

_____

_____

_____

_____

On Your Own

There are 54 Greek pots and 56 Roman pots in the museum. Each case can hold 9 pots. Which pots can be displayed in equal groups in the cases?

Write About It

How did you find the answer to On Your Own?

Objective 6.4: Find the basic facts involving division by 6, 7, 8, or 9.

Lesson 6-4: Divide by 6, 7, 8, or 9

1. What division problem does each set of MathFlaps show?

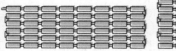 

_____ _____

2. Find all the numbers in the list that are divisible by 7. Write the facts.

14   24   27   28   30   42   45

_____

_____

3. Divide by 9. Fill in the table.

Dividing by 9

Dividend	Quotient	Remainder
10		
20		
30		
40		
50		
60		

4. Divide by 6. Fill in the table.

Dividing by 6

Dividend	Quotient	Remainder
21		
24		
27		
30		
33		

5. Solve.

a. 36 ÷ 6 = _____

b. 48 ÷ 6 = _____

c. 54 ÷ 6 = _____

d. 42 ÷ 6 = _____

6. Jessica drew 63 comic strips. Drea drew 64. Who will be able to make a book with exactly 8 comic strips on each page?

A Jessica   **B** Drea

C Both   **D** Neither

7. Mr. Ashman received some new T-shirts, and he displayed them in 7 equal stacks. How many T-shirts might he have received?

8. Look back at the dividends in exercise 3. Then look at the remainders you found when you divided by 9. What pattern do you see?

Objective 6.4: Find the basic facts involving division by 6, 7, 8, or 9.

Lesson 6-5: Relate Multiplication and Division

Activity 1

Use MathFlaps to show 4 × 3.

Then write the other facts in the family.

 4 groups of 3

3 groups of 4

Write both multiplication facts.
4 × 3 = 12   3 × 4 = 12

Now divide the total.
12 ÷ 4 = 3   12 ÷ 3 = 4

Practice 1

Use MathFlaps to show 8 × 3.

Then write the other facts in the family.

Write both multiplication facts.
_____   _____

Now divide the total.
_____   _____

Activity 2

Start with 21. How many groups of 7 can you make?

     3 groups

21 ÷ 7 = 3

Start with 21 counters. Make 3 groups. How many are in each group? 7

21 ÷ 3 = 7

Make 7 groups of 3. What is the total? 21

3 × 7 = 21   7 × 3 = 21

Practice 2

Start with 24. How many groups of 8 can you make? _____

Write all four facts your model shows.

_____ × _____ = _____

_____ × _____ = _____

_____ ÷ _____ = _____

_____ ÷ _____ = _____

On Your Own

Make a model to solve: 42 ÷ 6 = _____.

List all four facts your model shows.

_____   _____

_____   _____

Write About It

Explain how one model can show 4 related facts.

Objective 6.5: Write multiplication and division fact families.

Lesson 6-5: Relate Multiplication and Division — Understand It

Example 1

Write the fact family for 4, 6, and 24. Circle the total in each fact.

4 × 6 = (24)

6 × 4 = (24)

(24) ÷ 4 = 6

(24) ÷ 6 = 4

Practice 1

Write the product of 7 and 9 in the circle.
Complete the fact family.

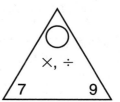

7 × 9 = ○

____ × ____ = ○

○ ÷ ____ = ____

○ ÷ ____ = ____

Example 2

Use a related fact to solve 32 ÷ 4 = ____.
Start with the total and divide.

32 ÷ ____ = 4

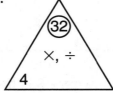

Multiply. The total should be the product.

____ × 4 = 32

4 × ____ = 32

Now complete one fact to find the missing number.
4 × 8 = 32, so 8 is the missing number.
32 ÷ 4 = 8

Practice 2

Use a related fact to solve 48 ÷ 6 = ____.
Start with the total and divide.

_____

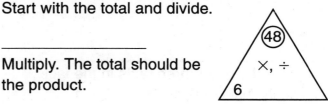

Multiply. The total should be the product.

_____

_____

Now complete one fact.
Write the missing number in the triangle.

48 ÷ 6 = ____

On Your Own

Complete the triangle. Write the fact family.

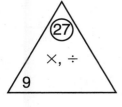

_____   _____

_____   _____

Write About It

Explain how to use the product in a fact family.

Objective 6.5: Write multiplication and division fact families.

Lesson 6-5: Relate Multiplication and Division

Try It

1. Which number belongs in a fact family that contains 2 and 18? Circle the letter of the correct answer.

   A 6　　　　B 7

   C 8　　　　D 9

2. Write the fact family that contains the product 72 and the factor 9.

   _____　　_____

   _____　　_____

3. Explain how 4 × 10 and 40 ÷ 4 have a similar meaning. Talk about equal groups in your answer.

4. Make a model to solve 28 ÷ 4. Write all the facts that your model shows.

   _____　　_____

   _____　　_____

5. Write the other three facts in the fact family that contains 18 ÷ 18 = 1.

6. Write the fact family that contains 2 and 8, where 8 is the dividend.

7. Complete the triangle. Write the fact family.

   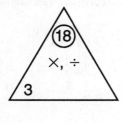

8. A fact family has two identical numbers. The dividend is between 30 and 40. Write the fact family.

Objective 6.5: Write multiplication and division fact families.

Topic 6 — Division Facts

Topic Summary

Circle the letter of the correct answer. Explain how you decided.

1. Rob rode his bike 42 miles in a week. If he rode the same distance each day, how far did he ride each day?

   A 42 miles

   B 6 miles

   C 294 miles

   D 8 miles

   _____
   _____
   _____
   _____

2. Which of these does **not** belong in the same fact family as 56 ÷ 8 = 7?

   A 8 × 7 = 56

   B 7 × 8 = 56

   C 7 + 8 = 15

   D 56 ÷ 7 = 8

   _____
   _____
   _____
   _____

Objective: Review division skills.

Topic 6 — Division Facts

Mixed Review

1. Find 45 ÷ 5 and show how to check using multiplication.

   _____

   _____

   Volume 2, Lesson 5-2

2. Find the missing addend.

   a. 8 + _____ = 15

   b. 4 + _____ = 13

   c. _____ + 9 = 15

   d. 9 + _____ = 19

   Volume 2, Lesson 4-5

3. Find each sum.

   a. 3 + 6 + 8 _____

   b. 5 + 4 + 9 _____

   c. 8 + 2 + 5 _____

   d. 6 + 6 + 9 _____

   Volume 2, Lesson 4-4

4. Which of the following is less than 98? Circle the letter of the correct answer.

   A 98   B 809

   C 908   D 89

   Volume 1, Lesson 2-3

5. Fill in the blanks.
   4,903

   a. There are _____ hundreds.

   b. There are 3 _____.

   c. 0 is in the _____ place.

   d. The value of the 4 is _____.

   Volume 1, Lesson 3-1

6. Round 835,721 to the nearest

   a. hundred thousand. _____

   b. ten thousand. _____

   c. thousand. _____

   Volume 1, Lesson 3-4

Objective: Maintain concepts and skills.

Words to Know/Glossary

A
addend — A number that is added to another number.

addition — Joining groups or increasing a quantity.

array — A picture that shows a multiplication fact.

associative property of multiplication — You can group factors in any way, and the product will be the same. 3 × (2 × 4) = 24 and (3 × 2) × 4 = 24.

C
commutative property of multiplication — You can multiply factors in any order, and the product will be the same. 3 × 8 = 24 and 8 × 3 = 24.

D
dividend — The number you divide.

divisible — A number is divisible by another number when there is no remainder after division.

divisor — The number you divide by.

F
factor — A number that you multiply by another number to equal a product.

M
multiplication — A way to join groups of equal size.

Multiplication Facts Table — The rows and columns in the multiplication facts table can help you find the facts.

N
number line — A line that shows numbers from smallest to greatest.

P
product — The answer to a multiplication problem.

Q
quotient — The answer to a division problem.

R
remainder — The number left over after you divide.

repeated addition — A way to multiply by adding the same number as many times as needed.

S
skip count — Use a counting pattern to multiply.

subtraction — Taking away, comparing, or decreasing a quantity.

Word	My Definition	My Notes

Word	My Definition	My Notes

Index

A
addition
    strategies, 8–10
    and subtraction, 14–16
    of three 1-digit numbers, 11–13
addition facts, 14–16
associative property
    of addition, 5–7
    of multiplication, 32–34

C
choosing operations to solve problems, 2–4
commutative property
    of addition, 5–7
    of multiplication, 32–34

D
dividend, 44–46, 51–52, 55
divisible, 51, 54
division
    by 2, 3, 4, or 5, 50–52
    by 6, 7, 8, or 9, 53–55
    families of facts, 56–58
    meaning of, 44–46
divisor, 44–46
doubles, 36

F
factors, 30
families of facts, 20–22
    division and multiplication, 56–58

M
Mixed Review
    4: Addition and Subtraction Facts, 24
    5: Multiplication Facts, 42
    6: Division Facts, 60
multiples, 35
multiplication
    by 2, 5, and 10, 29–31
    facts, 38–40
    families of facts, 56–58
    meaning of, 26–28
    patterns and properties of, 32–34
    strategies, 35–37
multiplication facts table, 29, 38–40

O
one, 47–49
operations
    choosing to solve problems, 2–4

P
patterns and properties, 5–7
products, 30
properties, 5–7
    of multiplication, 32–34
    of zero and one, 47–49
    See associative property; commutative property

Q
quotient, 44–46, 51–52, 55

R
relating addition and subtraction, 14–16
remainder, 50–53, 55
repeated addition, 26

S
skip count, 27
strategies
    of addition, 8–10
    of multiplication, 35–37
    of subtraction, 17–19
subtraction
    and addition, 14–16
    strategies of, 17–19

T
Topic Summary
    4: Addition and Subtraction Facts, 23
    5: Multiplication Facts, 41
    6: Division Facts, 59

Z
zero, 47–49